When you look in your backpack, what do you find?

Chances are that almost everything in there can be traced back to the Earth—the *soil*, to be more precise.

Your t-shirt is made of cotton that grew in soil of the southern U.S. Your wooden pencils are made from cedar trees harvested from well-drained forest soil. Your soccer ball is leather that came from cows that grazed on fertile prairies. The computer chip in your game controller contains silicon processed from our planet's plentiful sands. And that half-eaten apple, it once hung from a tree with roots stretching deep into the soil. Even the water you drank with lunch was purified as it passed through many layers of soil.

Soil is so much more than dirt. It supports us, and without it we could not survive. Soil is life.

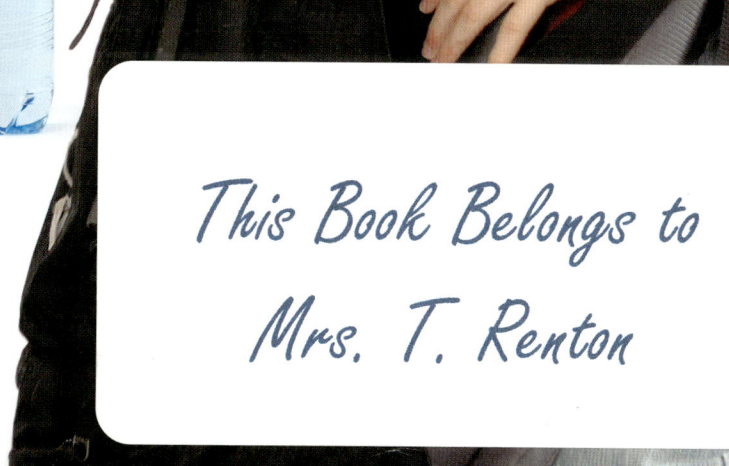

This Book Belongs to
Mrs. T. Renton

Dig These Words

Amino acids – Molecules that join together to form proteins, the building blocks of all organisms.

Anaerobic – Without oxygen as a part of the environment. The opposite of aerobic.

Antibiotic – A chemical that weakens or stops bacteria, fungi, and other microorganisms from growing.

Archaeologist – A scientist who studies ancient human cultures.

Arid – Dry, like a desert.

Bedrock – Mass of solid rock beneath the soil. Can be parent material if it is close enough to the surface to weather into soil.

Blocky – A soil structure. Soil particles are arranged into shapes that resemble small cubes with sharp or rounded edges.

Clay – The smallest-sized soil particles. Often have plate-like shapes. Feels sticky when wet. Also refers to a soil texture that consists of at least 40% clay particles.

CLORPT – The five factors that influence what type of soil forms: **cl**imate, **o**rganisms, **r**elief (landscape), **p**arent material, and **t**ime.

Columnar – A soil structure. Soil particles are arranged into tall vertical shapes or columns often with rounded tops.

Compost – The remains of plants and animals after they have decomposed. Can be used to fertilize soil and to improve its structure and ability to hold water.

Compounds – The combination of two or more elements. For example, hydrogen and oxygen combine to make water.

Decompose (Decomposition) – To break down a compound into simpler compounds. Often accomplished with the help of micro-organisms.

Developed soil – A soil that has had a long time to form, such as most tropical soils. A mature soil.

Deficiency – Lacking in something important. A deficiency of nutrients in a plant, for example, can stunt its growth.

Dormant – A state in which a plant or animal is not growing.

Ecologist – A scientist who studies interactions between organisms and their environment.

Eluviated horizon (E horizon) – A horizon from which minerals, clay, and/or organic matter have been leached.

Enzyme – A protein that increases the rate of chemical reactions in an organism's cells.

Erode (Erosion) – To wear away, or remove, rock or soil particles by water, ice, and/or gravity.

Estuary – A semi-enclosed body of water with a source of fresh water and an outlet to the ocean.

Fertility – The ability of a soil to supply essential nutrients to plants.

Fertilizer – A substance added to soil that contains plant nutrients such as nitrogen, phosphorus, and potassium.

Fungus – A kingdom of organisms distinct from animals and plants. Most fungi get their energy from decomposing plants and animals.

Granular – A soil structure. Soil particles are arranged into shapes that resemble granola. Has lots of pores.

Groundwater – Water that collects underground in the pore spaces of soil and rock. An important source of drinking water.

Horizon – A layer of soil with properties that differ from the layers above or below it.

Humus – Organic matter such as highly decomposed leaves.

Leaching – The removal of minerals and nutrients from a soil or a horizon as water passes through it.

Loam – A soil texture with moderate amounts of sand, silt, and clay, sometimes in nearly equal proportions. Good texture for farming and gardening.

Macronutrients – Nutrients needed by organisms in relatively large quantities.

Massive – A soil that has no structure. Soil particles are completely stuck together.

Microbes – Microscopic organisms, such as bacteria and fungi. Microbes represent the most abundant soil organisms.

Microbiologist – A scientist who studies microscopic organisms, or microbes.

Minerals – The inorganic particles in soils that weather from rocks.

Mottles (Mottling) – Spots or blotches of color(s) in a soil that differ from that soil's dominant color.

Nitrogen (N) – Macronutrient essential to living things like plant growth and building proteins. Often added to agricultural and garden soils.

Nutrients – Elements or compounds that nourish organisms. Essential for growth and reproduction.

Organic matter – Material derived from the decay of plants and animals. Always contains compounds of carbon and hydrogen.

Organisms – Living things such as bacteria, fungi, plants, or animals.

Parent material – The material from which a soil formed. Can be bedrock or materials carried and deposited by wind, water, glaciers, and/or gravity. The C horizon in a soil profile.

Peat – Partially decayed organic matter that accumulates in environments that stay wet.

Ped – The structural unit formed when soil particles (sand, silt, and clay) bind together.

Pedologist – A scientist who studies soils.

Perennials – Plants that live for more than two years as opposed to annuals that grow each year from seeds or biennials that live for only two years.

By David L. Lindbo, Wendy Greenberg, Laurel Hartley, John Havlin, Thomas E. Loynachan, Monday Mbila, Bianca Moebius-Clune, Emily Stockman, Janis Boettinger, Mary E. Collins, John Omueti, Dennis Osborne, Deborah A. Kozlowski, Chien-Lu Ping, Ray Weil

Written with Judy Mannes

Designed by Katherine Lenard/FasterKitty LLC

Managing Editor: Lisa Al-Amoodi

SSSA K-12 Committee: Thomas Loynachan, Chair; Members: Vernon Cardwell, Mark Carleton, Cecil Dharmasri, Sherry Fulk-Bringman, Wendy Greenberg, Laurel Hartley, John Havlin, Margaret Holzer, David Lindbo, Monday Mbila, Bianca Moebius-Clune, John Omueti, Dennis Osborne, Ellen Phillips, Clay Robinson, Susan Schultz, Emily Stockman; Ex Officio Members: Susan Chapman, Paul Kamps

Soil Science Society of America
677 South Segoe Road, Madison WI 53711 USA
www.soils.org • 608-273-8080

The Soil Science Society of America (SSSA) is an international scientific society that fosters the transfer of knowledge and practices to sustain global soils. SSSA is the professional home for 6,000+ members dedicated to advancing the field of soil science. It provides information about soils in relation to crop production, environmental quality, ecosystem sustainability, bioremediation, waste management, recycling, and wise land use.

ISBN: 978-0-89118-848-3

Library of Congress Control Number: 2008930771

Printed in the United States of America.

10 9 8 7 6 5 4 3 2

Permafrost – A soil horizon, or layer, that remains frozen year round.

Phosphorus (P) – Macronutrient essential to all living things like flowers, fruits, seeds in plants, and the nervous system in animals. Often added to agricultural and garden soils.

Photosynthesis – The process by which plants, some bacteria, and some algae use sunlight to convert carbon dioxide and water into food and oxygen.

Platy – A soil structure. Soil particles are arranged into shapes that resemble flat plates.

Pores – The space between soil particles, which can be filled with water or air. A **porous** soil has lots of pores.

Potassium (K) – Macronutrient essential to all living things like water uptake and pest resistance in plants; muscles and blood circulation in animals. Often added to agricultural and garden soils.

Prismatic – A soil structure. Soil particles are arranged into shapes that resemble columns.

Productive – A term used to describe a soil that has the capacity to grow an abundance of crops.

Relief – The shape of the land surface created by features such as hills and valleys.

Runoff – Water from precipitation or irrigation that does not soak into the soil but flows off the land and reaches streams and rivers.

Salinization – The build-up of salts in soil. Often occurs in arid environments.

Sand – The largest-sized soil particles. Sand feels gritty. Also refers to a soil texture that consists of at least 85% sand particles.

Sediment – Any particle of soil or rock that has been deposited by water, wind, glaciers, or gravity.

Sewage – Waste that goes down a drain (such as those in bathrooms, kitchens, and laundry rooms) to a treatment plant or septic system.

Silt – Soil particles in between sand and clay in size. Silt feels like flour (smooth and velvety). Also refers to a soil texture that consists of at least 80% silt particles.

Single-grained – A soil that has no structure. Soil particles are not bound to each other in any way, such as beach sand.

Slope – A landscape, or surface, that is tilted or inclined.

Sludge – Semi-solid material left behind after sewage has been processed in a treatment plant. May be used as a fertilizer in some instances.

Sod – Grass and the soil beneath it, held together by roots. Can be cut into blocks and used as a building material.

Soil – A mixture of minerals, organic matter, water, and air, which forms on the land surface. Can support the growth of plants.

Soil profile – A section of the soil that has been cut vertically to expose all its horizons, or layers.

Soil structure – The arrangement of soil particles into clusters, called peds, of various shapes that resemble balls, blocks, columns, or plates.

Soil texture – The relative proportions of sand, silt, and clay particles.

Subsoil (B horizon) – The soil horizon rich in minerals that eluviated, or leached down, from the horizons above it. Not present in all soils.

Tissue – A group of cells in an organism that work together, such as muscles in an animal or the outer surface of leaves in a plant.

Topsoil (A horizon) – Mostly weathered minerals from parent material with a little organic matter added. The horizon that formed at the land surface.

Transform – To change from one thing into another or from one state into another, like a liquid into a gas.

Tropical – The area of land and ocean that lies between 23.5° north and south of the equator.

Tundra – An area in cold regions, such as in the arctic or on mountains, where the growing season is very short.

Uptake – The ability of a plant to absorb water and nutrients.

Weather (Weathering) – To break down rocks and minerals at or near Earth's surface into smaller particles and soil.

Wetland – An area of land where the soil is saturated with water, such as a marsh, swamp, or bog.

Table of Contents

Heaven is under our feet as well as over our heads.

Henry David Thoreau

Soil is not dirt

Dirt is what gets on our clothes or under our fingernails. Something to wash off, to get rid of. At a glance, dirt and soil may look the same, but there is a big difference.

So, what on Earth is soil?

It is a complex mix of ingredients: **minerals**, air, water, and **organic matter**—countless organisms and the decaying remains of once-living things. Soil is made of life. Soil makes life. And soil is life.

We want to keep and protect soil. Dirt is something we want to get rid of.

THE SKIN OF THE EARTH

Soil is the thin outermost layer of Earth's crust. Like our own skin, we can't live without soil. Why?

- Most of our food comes directly or indirectly from plants anchored in, and nourished by, soil.

- Part of the oxygen we breathe is produced by plants living in soil.

- Much of the water we drink and use every day soaked into, and was filtered by, soil.

- And nearly everything we build is built on soil, and often with it.

Like our skin, it is easy to take soil for granted— *and* to damage it.

Earth's radius is approximately 20,889,272 feet (6,367,050m)

Approximately 3–6 feet (0.91–1.83m)

In our solar system, Earth is a unique planet. It has vast oceans, abundant life, and a "breathable" atmosphere.

And, it has soil!

---

4 **Soil: Up Close and Personal** **Soil means different things depending on who you are.**

FARMER
It is where crops grow.

"Soil is at the base of everything we do: our food, our clothes, even fuel for our cars. It's my job to use the soil in a sustainable way to provide all these things."

ENGINEER
It is a foundation to build on.

"Soils are the backbone of the world we live in. But they range in strength from being able to support a skyscraper to not being able to support your own weight."

ARCHAEOLOGIST
It holds clues of past cultures.

"As an archaeology student, the first thing I learned to do was hold a trowel to dig in the soil. If you don't get that right you can never be an archaeologist."

Some of the first written words were recorded on clay tablets, such as this ancient Egyptian one.

Soil was so important to early civilizations that people worshiped soil gods and goddesses. Geb, here, was the Egyptian god of the Earth, always shown having a green, leaf-covered body.

We can even make links between proper names and the soil: "Adam" comes from the Hebrew word *adama*, which means soil, or earth.

For thousands of years people have shaped clay into urns and other vessels. Some, like this ancient Greek pitcher, were elaborately decorated with paints, sometimes made from soil.

For generations people have used soil to build their houses. Native Americans in the American Southwest dried the desert soil into adobe bricks.

Settlers on the Midwestern prairies cut **sod** into bricks to build their homes.

People of the Navajo Nation built earthen houses called hogans.

SOIL: WE'RE ROOTED IN IT

Did you know that "earth" and "soil" are some of the earliest words ever written? These words tell us a lot about the importance of soil to people. The Romans called Earth *solium*, from which we get the English word "soil." To the Greeks, soil, or Earth, was *pedon*, which is why we call some soil scientists "**pedologists**."

ECOLOGIST
It supports the web of life.

"Soil connects all life and all ecosystems. Even in the harshest ecosystems the soil is a refuge for life."

POTTER
It provides clay to work with.

"I love seeing the clay transformed into useful ceramic things, and the sand and minerals from the soil fired into shining glazes."

SOIL SCIENTIST
It's all of these things, but most of all . . .soil is life.

"There is more to soil than meets the foot!"

How do soils form?

Break it down . . . Build it up.

Dig down deep into any soil, and you'll see that it is made of layers, or **horizons**. Put the horizons together, and they form a **soil profile**. Like a biography, each profile tells a story about the life of a soil.

Every soil formed from **parent material,** a deposit at the Earth's surface. The material could have been **bedrock** that **weathered** in place or smaller materials carried by flooding rivers, moving glaciers, or blowing winds. Over time, sun, water, wind, ice—and living creatures—help **transform**, or change, the parent material into soil.

Soil ABCs

Each horizon in a soil profile is different from the one above it and the one below it. But learning horizons is as easy as learning your ABCs—just in a slightly different order. From top to bottom, the horizons are **O, A, E, B, C,** and **R.**

Try this memory aid: Old **A**unt **E**thel **B**akes **C**ookies **R**egularly. **Or you can invent your own.**

A vertical cut through the soil is called a soil profile. Every soil profile may not have all of the horizons shown here, nor are the boundaries between the layers always as sharply defined as the ones in this illustration.

O (humus or organic)
Mostly organic matter such as **decomposing** leaves. The **O** horizon is thin in some soils, thick in others, and missing in yet others.

A (topsoil)
Mostly minerals from parent material with a little organic matter added. A good material for plants and other organisms. You can find lots of roots here.

E (eluviated horizon)
Leached of clay, minerals, and organic matter, which makes the **E** horizon sandier and lighter in color than the **A** horizon above and the **B** horizon below it. Often found in some older soils and forest soils.

B (subsoil)
Rich in minerals that **leached** (moved down) from the **A** or **E** horizons and accumulated here. Not present in all soils.

C (parent material)
The deposit at Earth's surface from which the soil developed.

R (bedrock)
A mass of rock that forms the parent material for some soils if the bedrock is close enough to the surface to weather.

ADDITIONS

Rain adds water. Dust adds minerals. Animal wastes add organic matter and *nutrients*. Humans add *fertilizers*.

LOSSES

Water evaporates into the air. Nutrients are taken up by plants. Soil particles wash away in a storm. Organic matter may decompose into carbon dioxide. Minerals and nutrients leach into *groundwater*.

TRANSLOCATIONS
(MOVEMENT WITHIN THE SOIL)

Gravity pulls water down from top to bottom. Evaporating water draws minerals up from bottom to top. Organisms carry materials every which way!

TRANSFORMATIONS
(ONE COMPONENT CHANGES INTO ANOTHER)

Dead leaves decompose into *humus*. Hard rock weathers into soft *clay*. Oxygen reacts with iron, "rusting" the soil to a reddish color.

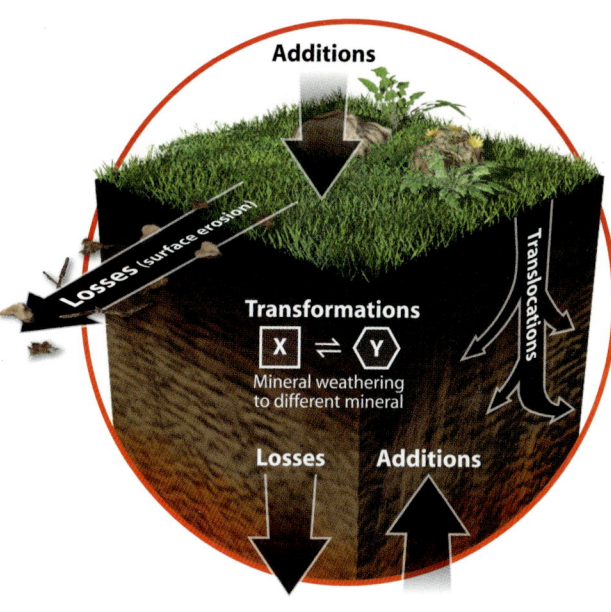

LOOKS CHANGE WITH AGE

As a soil ages, it gradually starts to look different from its parent material. That's because soil is dynamic. Its components—minerals, water, air, organic matter, and **organisms**—constantly change. Some components are added. Some are lost. Some move from place to place within the soil. And some components are totally changed, or transformed.

A Snapshot in Time   Soil horizons take hundreds to thousands of years to form.

A soil profile is like a snapshot, capturing what the soil looks like now. In the past, it looked different and will look different in the future.

AGE IN YEARS

0   10   100   1,000   10,000   100,000

A - topsoil
E – eluviated horizon
B – subsoil
C - parent material
R – bedrock

>>Inside Scoop

The word "human" is probably related to the word "humus," Latin for "earth" or "ground." Soil scientists use the word "humus" to mean a soil's decomposed organic matter.

CLORPT for short

Soils differ from one part of the world to another, even from one part of a backyard to another.

They differ because of where and how they formed. And over time, five major factors control how a soil forms. These factors are **c**limate, **o**rganisms, **r**elief (landscape), **p**arent material, and **t**ime. That is *CLORPT* for short!

Each soil has its own history.

Charles Kellogg, Former Director, Soil Survey Division, USDA

>>Inside Scoop

Soil is slow "growing." A single inch of soil can take hundreds to thousands of years to develop.

CLimate

Temperature and moisture influence the speed of chemical reactions, which in turn help control how fast rocks weather and dead organisms decompose.

Soils develop fastest in warm, moist climates and slowest in cold or *arid* ones.

TROPICAL   DESERT

Organisms

Plant roots spread, animals burrow, and bacteria eat. These and other soil organisms speed up the breakdown of large soil particles into smaller ones.

Roots are a powerful soil-forming force, cracking rocks as they grow. And roots produce carbon dioxide that mixes with water and forms an acid that wears away rock.

Relief

The shape of the land and the direction it faces make a difference in how much sunlight the soil gets and how much water it keeps.

Deeper soils form at the bottom of a hill than at the top because gravity and water move soil particles down the **slope**.

Parent material

Just like you inherited some characteristics from your parents, every soil inherited traits from the material from which it formed.

Soils that form from limestone bedrock, for example, are rich in calcium. Soils that formed from materials at the bottom of lakes are high in clay.

Time

Older soils differ from younger soils because they have had longer to develop.

In the northern U.S., soils tend to be young because glaciers covered the surface during the last Ice Age, which kept soils from forming. In the southern U.S., there were no glaciers. There, the soils have been exposed for a longer time, so they are more weathered.

I ❤ SOIL

**Emily K. D. Stockman
Soil Scientist, Farmer, Teacher
Massachusetts**

*I live on a moderately well-drained **loam** soil in Plainfield, Massachusetts, with my husband Ty and my son Henry. I began my soils career at an early age as mud pie maker extraordinaire. Keeping my hands dirty, I spent a number of years as an organic farmer. I received my B.S. in soil science and my M.S. in **wetland** soil science from the University of Massachusetts Amherst.*

I am currently a consulting soil scientist specializing in wetlands, and I travel all over Massachusetts. Even though my state's soils are young (only about 10,000 years old!), they have been affected by CLORPT. In low-lying areas, I collect data on the organic-rich, gray soils of wetlands. In upland areas, I examine farmed soils. There, I find earthworms and crop roots thriving in the yellow, oxygen-rich B horizon. To me, digging a soil pit is like opening a present—each present has diverse colors, different layers, and a new story to tell.

Sizing up soils

What makes soil *soil*?

Have you ever made a "pie" with mud? Filled a bucket with sand? Each one is a soil—a complex mixture of solids, liquids, and gases. Soils can be described using three characteristics: texture, structure, and color.

TEXTURE: What are the percentages of sand, silt, and clay particles that make up the soil?

STRUCTURE: How do the particles fit together?

COLOR: Is the soil light or dark? Is the color even (uniform) or spotty (*mottled*)?

SOIL HAS TEXTURE

The particles that make up soil are categorized into three groups by size—*sand*, *silt*, and *clay*. Sand particles are the largest and clay particles the smallest. Although a soil could be all sand, all clay, or all silt, that is rare. Instead most soils are a combination of the three.

The relative percentages of sand, silt, and clay are what give soil its *texture*. A loamy texture soil, for example, has nearly equal parts of sand, silt, and clay.

There are 12 soil textures represented on this triangle. Soil scientists use this device so that terms like "*clay*" or "*loam*" always have the same meaning. Each texture corresponds to specific percentages of sand, silt, or clay.

You could compare the relative sizes of soil particles to a basketball (sand), a softball (silt) and a BB (clay).

SILT – 0.05 to 0.002 mm

SAND – 2.0 to 0.05 mm

CLAY – less than 0.002 mm

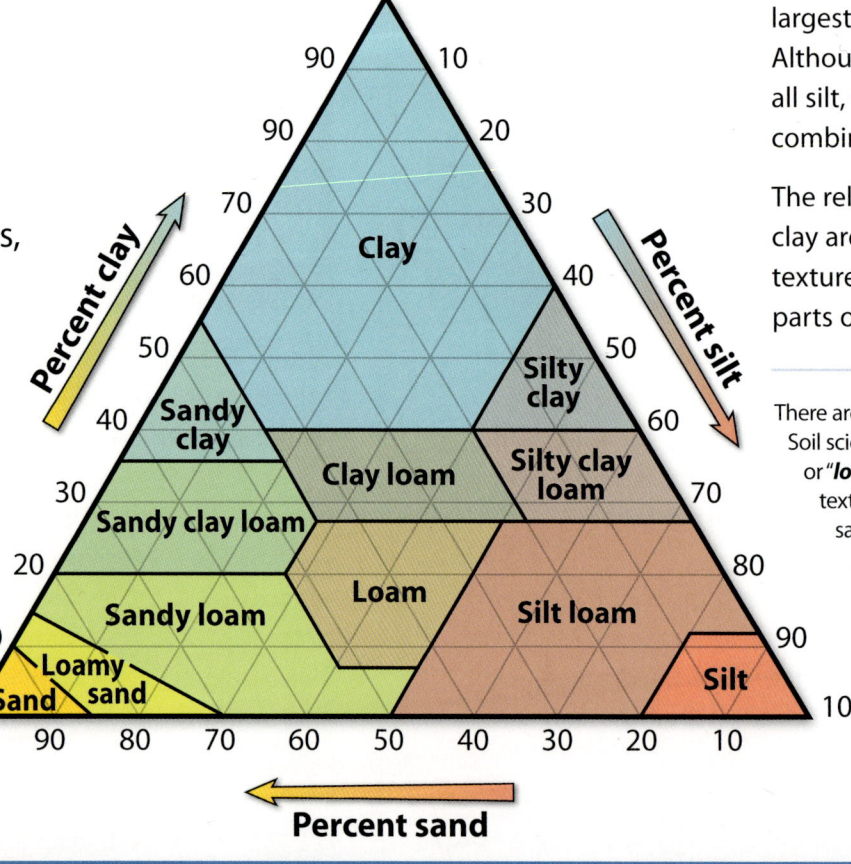

Percent clay

Percent silt

Percent sand

100 · 90 · 80 · 70 · 60 · 50 · 40 · 30 · 20 · 10 · 100

Clay

Sandy clay

Silty clay

Clay loam

Silty clay loam

Sandy clay loam

Loam

Silt loam

Sandy loam

Loamy sand

Sand

Silt

Get to Know Your Soil . . . Just like a soil scientist would do.

Try these tests to determine your soil's texture, and its sand, silt, and clay content.

FEEL IT: Dry and crush a small amount of soil. Rub it between your thumb and fingers. The grittier it feels, the higher the sand content. If it feels like flour the silt content is higher.

FORM A BALL: Squeeze moist soil in your hand. Can you form it into a ball? Pass the ball from hand to hand. The stickier it is and the longer the ball stays together, the higher the silt and clay content.

MAKE A WORM: Roll a handful of moist soil into a worm shape. The longer and thinner you can make it, the higher the clay content. Squeeze it into a thin ribbon. The more the ribbon flakes, the higher the silt content.

SOIL HAS STRUCTURE

Soil structure is the arrangement of soil particles into small clumps, called *peds*. Much like ingredients in a cake batter bind together to form a cake, soil particles (sand, silt, clay, and even organic matter) bind together to form peds. Peds have various shapes depending on their "ingredients" *and* on the conditions in which the peds formed: getting wet and drying out or freezing and thawing—or even people walking on or farming the soil.

Ped shapes roughly resemble balls, blocks, columns, and plates. Between the peds are spaces, or *pores*, in which air, water, and organisms can move. The sizes of pores and their shapes vary from soil structure to *soil structure*.

>>Inside Scoop

It's fun, but riding off-road can be tough on soil—compacting it, and killing plants that prevent erosion.

GRANULAR

Resembles cookie crumbs or granola with lots of pores (like good gardening soil).

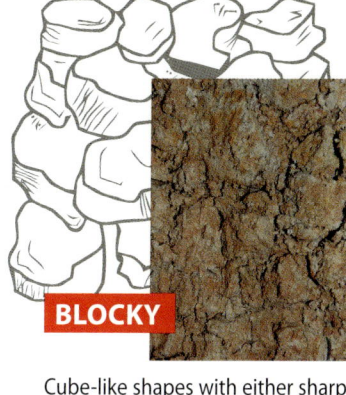

BLOCKY

Cube-like shapes with either sharp or smooth edges. Lots of pores.

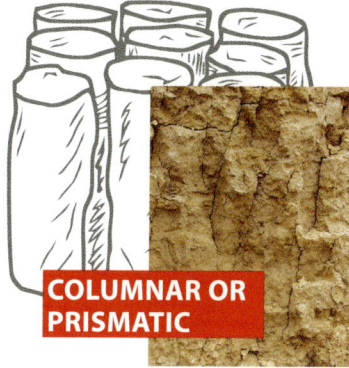

COLUMNAR OR PRISMATIC

Tall vertical shapes. Fewer pores than a blocky structure.

PLATY

Flat horizontal plates.

MASSIVE

No structure, no pores. Particles completely stuck together in large clumps (like modeling clay).

SINGLE-GRAINED

No peds. Particles stay separate (like beach sand).

Keeping Soils Healthy   If we treat soils right, we can keep them healthy.   11

STRUCTURE = HEALTH

A soil is said to be healthy when it has good structure. Plants grow best in a "well-structured" soil—a *porous* soil with lots of spaces for air, water, and nutrients to circulate. *But* plowing or driving over a soil too much can compact it. Its pores are squeezed together. The soil loses its structure, it becomes *massive*, and plants will not grow as well.

KEEP IT COVERED

The cornfield on the left has been plowed often. That disturbs the plant remains (roots, stubble) of the harvested crop, and exposes the soil to wind and rain, which may erode the soil.

The field on the right has been plowed rarely. The plant remains of the harvested crop stay in or on the ground, which add organic matter to the soil, help keep the structure loose, and prevent the soil from *eroding*.

TEXTURE + STRUCTURE =
SOIL BEHAVIOR

A soil's texture and structure can tell us a lot about how a soil will behave. *Granular* soils with a loamy texture make the best farmland because they hold water and nutrients well. *Single-grained* soils with a sandy texture don't make good farmland because water drains out too fast. *Platy* soils, regardless of texture, cause water to pond on the soil's surface.

I ♥ SOIL

**Deborah Kozlowski
Artist, Forester, Soil
Scientist, Teacher
North Carolina**

I've always loved nature, science, and art. And in college at the University of New Hampshire, I had a professor who demonstrated how soil was an important part of all those things. I went on to graduate school in soils at the University of Massachusetts, where one day I went to an ice cream parlor. On the wall was a beautiful tile mural of cows in a pasture. I said to myself, "I want to make things like that!"

Since then I have made thousands of tiles. I love seeing the clay from the soil transformed into useful ceramic things, and the sand and minerals from the soil fired into shining glazes.

A clay soil gets sticky, slick, *and* very messy when wet. And if its structure is platy, little water can soak into it. Water can also flow between flat peds and particles, which can cause you to slip and slide on a platy, clay soil.

A sandy soil is loose and easily moved around by wind, water, *and* people. That's why it is a good idea to stay off of sand dunes at the beach.

Understanding a soil's structure and texture is important to knowing where to build houses *and* roads to avoid soils likely to slide.

Soil Connects it All: "Dust" Across the Seas

CLOUDS

DUST

AFRICA

ATLANTIC OCEAN

This satellite image captures one of Africa's biggest exports—soil from the Sahara Desert. Whipped up by winds and carried into the atmosphere, fine particles of desert soil can travel thousands of miles across the Atlantic Ocean. The dust sometimes causes hazy skies in the southeast U.S. But the effects are not always bad: Saharan dust, for example, adds needed nutrients to the soils of the Amazon rainforests.

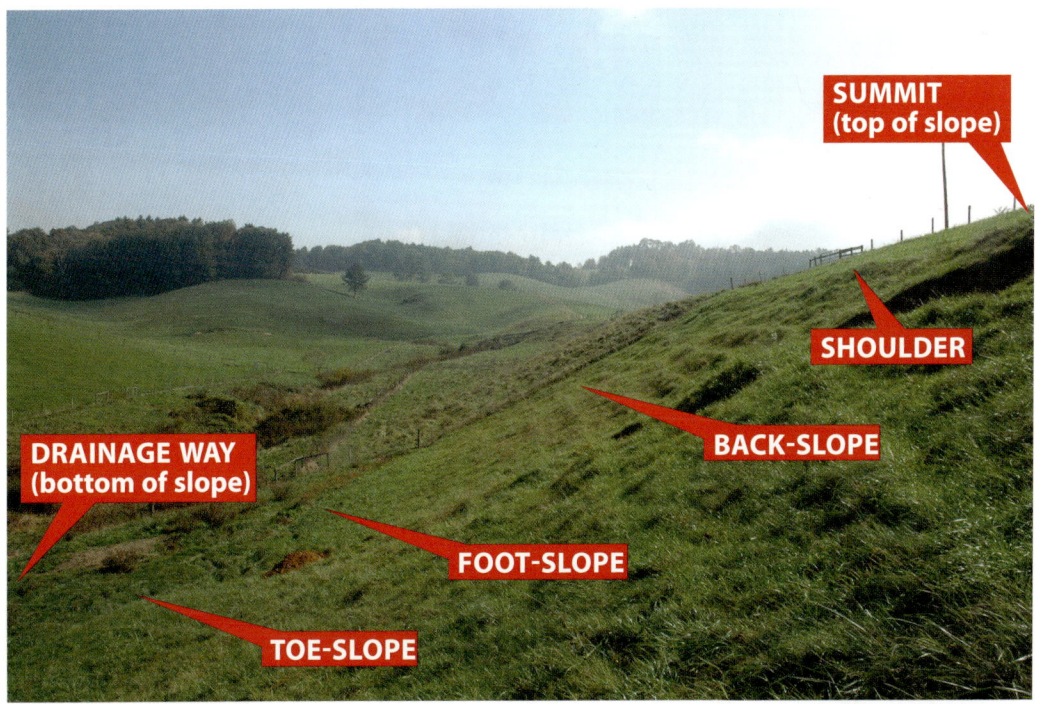

SUMMIT (top of slope)

SHOULDER

BACK-SLOPE

DRAINAGE WAY (bottom of slope)

FOOT-SLOPE

TOE-SLOPE

Soil color changes depending on where it is on a slope. At the top of the slope the soil drains well. At the bottom, water collects. Water removes minerals that coat the soil particles and color it. Colors change from bright red, to orange and yellow, and finally to gray as you move down the slope.

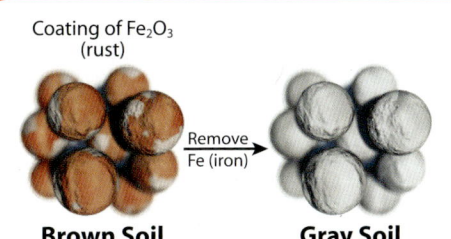

Coating of Fe₂O₃ (rust)

Remove Fe (iron) →

Brown Soil   **Gray Soil**

Iron oxide (rust) coats soil particles and turns them reddish brown. As water moves through the soil, the iron oxide is removed. The soil loses its color and turns from reddish brown to gray.

YOU CAN TELL A SOIL BY ITS COLOR

The color of a soil can tell us a lot about that soil.

Color can tell us about the soil's mineral content. Soils high in iron are deep orange-brown to yellowish-brown. And those with lots of organic material are dark brown or black. Organic matter masks all other coloring agents.

Color can also tell us about how a soil "behaves." A soil that drains well is brightly colored. One that is often wet and soggy has an uneven (***mottled***) pattern of grays, reds, and yellows.

DRAINAGE WAY

TOE-SLOPE

FOOT-SLOPE

SHOULDER OR BACK-SLOPE

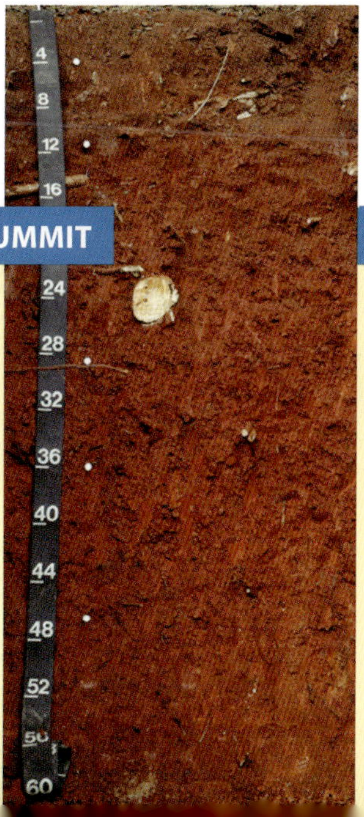

SUMMIT

Yikes, it's alive!

Watch your step. Soil swarms with life.

But most of it is out of your sight—in the ground or so small you need a magnifying glass or a microscope to see it. Many, many times more organisms live below ground than above it.

Soil provides a home for organisms, and in turn, organisms keep soil healthy. It's a partnership. Each sustains the other.

If you examine the upper few centimeters or inches of one square meter of ground (about 3 feet by 3 feet), you'll find millions of nematodes (tiny worms) in it for every bird or squirrel that you can see above the ground. This pyramid organizes diverse groups of soil organisms by their abundance. The smallest among them, it turns out, are the most abundant.

Vertebrates (1)
Snails and Slugs (100)
Potworms and Earthworms (3,000)
Insects and Spiders (5,000)
Rotifers (10,000)
Springtails (50,000)
Mites (100,000)
Nematodes (5,000,000)
Protozoa (10,000,000,000)
Fungi (100,000,000,000)
Bacteria (10,000,000,000,000)

Snails and slugs secrete a slimy coating that keeps their bodies from drying out as they slither across the soil.

Earthworms push their way through the soil by partially eating it. Their "manure" helps fertilize the soil. Earthworm poop has a granular structure.

In addition to spinning webs above ground, most **spiders** dig a burrow in the soil from which they also hunt.

Every big grazing mammal has a **dung beetle** associated with it that specializes in burying the grazer's droppings in the soil.

14

Under the microscope

By whirling hair-like cilia on their heads, **rotifers** swim through the water that collects on leaf litter and between soil particles.

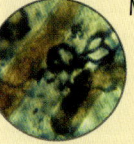
Springtails that live at the soil surface use an organ beneath their abdomen to "spring" away from predators. Those that live deep in the soil have no tail.

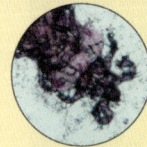
Mites chew plant debris into small pieces, which makes the debris available to yet smaller decomposers and *microbes*.

Much smaller, but more numerous and diverse than earthworms, **nematodes** live in every type of soil. Not picky eaters, they feed on roots, *fungi*, and most other soil organisms.

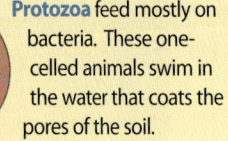
Protozoa feed mostly on bacteria. These one-celled animals swim in the water that coats the pores of the soil.

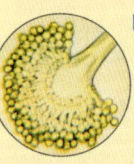
Fungi (like the common mushroom) recycle soil nutrients as well as serve as food for many soil animals. Some work with plants by giving them nutrients.

Plants could not live without the **bacteria** that collect around their roots. These microbes convert elements in the soil and air into nutrients that plants can absorb.

>>Inside Scoop

The microbe actinomycetes produces a chemical that gives Earth its "earthy" smell. It also enables camels to find water in the desert, as well as being a source of *antibiotics*.

New York skyscrapers rise above a forest soil.

The pyramids in Egypt rise above desert soils.

Where in the world of soils might you be, if you found . . .

. . . an old plow blade buried in it?

. . . a rubber boot stuck in it?

. . . an antler frozen in it?

. . . an acorn sprouting from it?

. . . an empty water bottle lying on it?

. . . a termite mound built from it?

Read on for clues.

Dig Deeper

Each state in the United States has an "official" state soil, like each has a state flower or bird. To find your state's soil, go to **http://soils.usda.gov/gallery/state_soils/**

INCEPTISOL
Slightly developed (young)

MOLLISOL
Deep, fertile

OXISOL
Very weathered

SPODOSOL
Sandy, acidic

ULTISOL
Weathered

VERTISOL
Shrinks and swells

prairie soils

Prairies are a type of grassland. They form in climates too dry to be a forest and too moist to be a desert. There are prairies in the Great Plains of North America, the Ukraine of Eastern Europe, and the Pampas of Argentina. Prairies often form in the interior of a continent.

Breakfast would be very different without prairie soils.

Prairie soils are rich, soft, and deep. They form under grasslands where the climate has warm summers and cold winters. When grassland plants die back in winter, their leaves and roots remain. The debris from the plants acts like mulch on a garden. It adds organic matter, which keeps the soil fertile.

MOLLISOL Latin *mollis* = soft

Prairie soils are Mollisols, which have a deep, dark layer of **topsoil** rich in organic matter. But even within a prairie, the soil can differ from region to region depending on the climate and the plants.

What differences can you see here?

MOLLISOL in a tall grass prairie

MOLLISOL in a short grass prairie

Prairie soils are very **productive**: They make up the world's great "breadbaskets," where most of the wheat, corn, soybeans, and other grains you probably ate for breakfast were grown.

Map legend:
- Short grass
- Tall grass
- Mixed grass

CANADA · UNITED STATES · MEXICO · Pacific · Atlantic

SHORT GRASS, TALL GRASS

In North America, the prairie is divided into three regions—tall grass, short grass, and mixed grass—because rainfall decreases from east to west.

The tallest grasses, which have the deepest roots, grow in the eastern prairie. Only short grasses can grow in the more arid western prairie. As a result, the soils form differently. Soils in the tall grass prairie have the darkest and thickest layer of organic matter.

Bison calf
Tens of millions of bison (buffalo) grazed on native grasses until farms, ranches, and towns replaced much of the prairie in the late 1800s.

PRAIRIE CLORPT

> CLimate

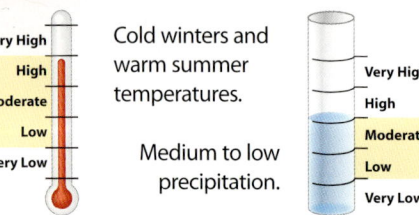

Cold winters and warm summer temperatures.

Medium to low precipitation.

Thermometer scales: Very High, High, Moderate, Low, Very Low

> Organisms

Bacteria, fungi, ground beetles, grasses, legumes (like purple clover), prairie dogs, bison (buffalo).

> Relief Rolling to relatively flat.

> Parent material

In North America, it is commonly windblown silt particles called loess. In the north, loess lies over rocks and other deposits left by glaciers. In the south, loess covers older sediments deposited by rivers.

> Time

Soils are older in the south (like Texas) than in the north (like Iowa), which was covered by glaciers until about 10,000 to 15,000 years ago.

Prairie Shopping List

Everything in this cart is connected to prairie soil, and there's much more than food.

- Aspirin (corn starch)
- Cooking oil (canola)
- Dog food (corn, soybean, beef)
- Ketchup (sweetener from corn)
- Soymilk (soybean)
- Work gloves (leather from cattle)
- Soft drinks (sweetener from corn)
- Diapers (cellulose from corn)
- Cat litter (wheat stalks)

Cattle are valuable for more than meat and milk. Their manure, like that of pigs and poultry, makes a good fertilizer.

Manure is collected from livestock and put into holding areas. Periodically it is applied to the soil to fertilize crop fields.

RECYCLE IT!

Keeping Prairie Soil Fertile

A prairie recycles its own nutrients, which keeps the soil fertile. When prairie plants and animals die, they remain in place. They eventually decompose and add nutrients to the soil. In contrast, when farmers harvest their crops, most of the nutrients are removed from the soil. As a result, farmers have to fertilize their soils.

Some farmers make their own fertilizers, or **compost**, by recycling manure from livestock and mixing it with corn, wheat and cotton stalks, and other materials.

The material that is being spread on this field contains nitrogen, phosphorus, and potassium—**macronutrients** essential for plant growth.

I ❤ SOIL

**Laurel Hartley
Soil Scientist
Colorado**

I was more interested in reading than playing with soil as a child. Perhaps if I had seen a book about soil like this one, I might have become interested in soil earlier in life. As it was, I became interested in soil in graduate school. While earning a Ph.D. in ecology from Colorado State University, I studied the black-tailed prairie dog, a really neat burrowing animal that lives in North American prairies (on sandy loam soils, to be specific!). I quickly learned that to fully understand prairie dogs, I had to understand their relationship with the soil.

After completing my Ph.D., I moved to Michigan State University's W. K. Kellogg Biological Station where I coordinated a program for K-12 teachers and also learned about greenhouse gas emissions from agricultural soils. I am currently an educator and researcher at the University of Colorado, Denver, studying animal-plant-soil interactions and science education.

tundra soils

Bring warm boots in winter and bug spray in summer.

Tundra soils form in very cold climates—at high latitudes, as in the arctic, and at high altitudes, as on mountains. In many locations, a permanently frozen layer, called permafrost, remains all year. But winter snow and ice and the frozen upper layer of soil above the permafrost melt during the tundra's brief summers. The ground gets very soggy, making it a perfect place for mosquitoes, flies, and other biting insects to breed. Insects, however, can't spoil the beauty of the vast wetlands common on tundra soils.

GELISOL Latin *gelare* = to freeze (think "gelato")

Tundra soils are Gelisols. A Gelisol contains permafrost, a layer of soil near the surface that is frozen year-round. Gelisols are usually black or dark brown because organic material accumulates in the upper layer (O horizon) and possibly in lower horizons.

Look for the permafrost layer in the soil profile.

GELISOL

ICE LENS

Water freezes in the subsurface of the Gelisol, creating a pocket, or lens, of ice.

STUMBLING OR TILTED FOREST

Trees that grow in tundra soils often tilt at various angles. As the tundra repeatedly freezes and thaws, the soil heaves and buckles, which make the trees stagger. Also, hard *permafrost* prevents trees from growing deep roots to stabilize them.

>>Inside Scoop

When permafrost melts, organic materials decompose and release carbon dioxide and methane into the atmosphere. There, these "greenhouse" gases trap heat and raise global temperatures.

TUNDRA CLORPT

> CLimate

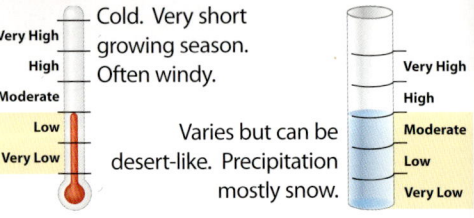

Cold. Very short growing season. Often windy.

Very High
High
Moderate
Low
Very Low

Varies but can be desert-like. Precipitation mostly snow.

Very High
High
Moderate
Low
Very Low

> Organisms

Anaerobic bacteria, fungi, lichens, mosses, sedges, shrubs (blueberry, crowberry, Labrador tea, arctic willow, dwarf birch), lemmings, caribou, geese and other migratory birds.

> Relief Varies. In the Arctic, tundra is relatively flat or rolling. On mountains, it is on gentle to moderate slopes.

> Parent material

Varies but often from loess (wind blown silt), weathered rocks, mixed silt and rock fragments, and sediments left behind by glaciers. Also organic matter.

> Time

Young at high latitudes, because the ground was covered by glaciers until 10,000 years ago. Can be old at high elevations and areas not covered by glaciers.

Patterned ground forms in tundra. Water fills cracks in the soil, freezes, and shapes the soil into polygons—a geometric shape with at least three sides. A polygon can range from one to 30 meters wide (the length of a yardstick to two school buses parked end to end).

Tundra Textures

From the air, tundra may look flat and boring. But if you walked across it, you would soon find out it is anything but. The ground is a jumble of cracks and gullies, mounds and hills, ponds and sinks—all created by permafrost.

As water in the soil freezes, it expands and reshapes the landscape. Sometimes tundra-like landforms can be found in areas that no longer have permafrost, like some mountains of New Hampshire. These are reminders that the climate has changed over time.

Pingo is an Eskimo word for a cone-shaped hill of soil. Pingos form when an underground pond of water freezes, expands, and pushes the soil upward. Pingos can rise as tall as a 20-story building and spread up to two kilometers (1.2 miles) in diameter.

KEEP IT COLD

Building on permafrost is tricky.

Permafrost is mostly frozen water in which ice crystals help hold the soil together. If the permafrost thaws, the ground can shift. Roads often buckle. House foundations sink. And pipelines risk collapsing. Building on permafrost—as on any unstable soil—requires special engineering.

Repeated freezing and thawing of the soil causes roads to heave, increasing maintenance costs. In Alaska, replacing 1 kilometer (0.6 mile) of road can cost up to $1.5 million.

An Alaskan homeowner built a new house to replace the one claimed by sinking permafrost. The air space under the new house will circulate cold air to keep the permafrost from melting and to keep the soil more stable.

I ♥ SOIL

**Chien-Lu Ping
Soil Scientist
Alaska**

I like to dig in soils because I have been interested in growing things and playing with "mud" since I was a kid. But I decided to become a soil scientist when I volunteered with a soil survey team in college. In the field, I was intrigued by the "looks" of different soil profiles under different vegetation and landscape positions. I currently teach soil science at the University of Alaska Fairbanks.

In the early years, I focused on soils in southern Alaska that formed in volcanic ash. My current focus is tundra soils, which have permafrost within 60 cm (24 in) of the surface. Soils with permafrost occupy less than 13% of the world's land surface area, but they store more than 30% of the world's terrestrial carbon. But these areas are shrinking due to global warming. Thus, tundra soils hold a key role in the global carbon cycle.

23

desert soils

Not all deserts are sandy—or blistering hot.

But all deserts are dry. They form in regions where the moisture from rain or snow evaporates faster than it can be replenished. But desert soils vary from place to place. Those that developed from sediments left by rivers or lakes can be deep. Others that formed on bedrock can be shallow.

The lack of moisture keeps some minerals from leaching out of the soil. Instead, the minerals accumulate and create cement-like soil horizons near the surface.

ARIDISOL

ARIDISOL Latin *aridus* = arid, or dry

Desert soils are dry for extended periods of time. Even when they receive rain or snow, the high rate of evaporation makes them dry out quickly. Desert soils typically are light in color because there is little vegetation to add organic material (which is dark). Also, the lack of moisture keeps microbes from rapidly decomposing plants when they die.

SOILS WITH A CRUST

Crusts often form on desert soils. Some desert crusts are biological—formed by communities of organisms such as bacteria, algae, mosses, and lichens. Others developed from physical processes. The impact of raindrops, for example, can create crusts with plate-like shapes.

PLATY CRUSTS

LIVING CRUSTS

>>Inside Scoop

The entire continent of Antarctica is a vast cold desert.

DESERT CLORPT

> CLimate

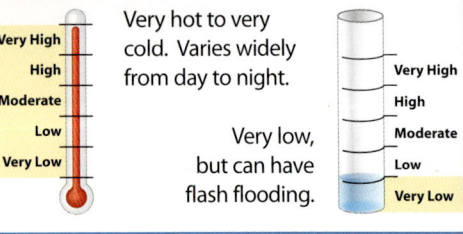

Very hot to very cold. Varies widely from day to night.

Very low, but can have flash flooding.

Very High / High / Moderate / Low / Very Low

Very High / High / Moderate / Low / Very Low

> Organisms

Vary by climate, but all are adapted to dry conditions. Bacteria, lichens, mosses, ants, termites, shrubs (sagebrush, creosote bush, saltbush), rodents, reptiles.

> Relief
Varies from nearly level to steep.

> Parent material

Varies: clays from ancient lakes, stream deposits, windblown sand.

> Time

Hundreds of thousands of years old to very young.

Salt Can Kill

By 3000 BC, the Sumerians built large cities in the deserts of southern Mesopotamia (now mostly Iraq). Using irrigation, they farmed the desert soils and created large food surpluses that made their civilization possible. But around 2200 BC, the civilization collapsed. Scientists debate why, but one reason may be tied to the soil. Irrigating in dry climates can cause a buildup of salt, a process called salinization. Few crops can tolerate salt.

This is all that is left of Ur, one of eight large Sumerian cities. No large urban societies have developed in southeastern Iraq since that time. The soil remains too salty to grow crops.

This irrigated farm field is suffering from *salinization*. All water contains salts dissolved in it. And when water evaporates, the salts stay behind as solids. Irrigation, however, adds much more water to the soil than nature would. So more and more salts build up over time, especially in dry climates where water evaporates quickly.

Just a few centimeters of rain can trigger a desert bloom—waking up wildflower seeds that were lying **dormant** on the desert floor for years. Rain also encourages **perennials** like cactus to flower.

The water in this irrigation canal came from the Colorado River, which supplies water to most of Southern California's cities and suburbs as well as its farms. As urban and suburban populations grow, competition for water is also increasing in this dry region.

An irrigated farm in California's Central Valley.

MAKING THE DESERT BLOOM

Desert soils can be fertile, but often too dry to farm—unless they are irrigated. With irrigation, for example, the arid soils of California's Central Valley produce more than 250 types of fruits and vegetables. Irrigation water is either taken from rivers or streams, or aquifers (underground areas of porous rock that hold lots of water).

But sometimes, irrigation can use up water sources faster than they can be recharged by snow and rain. Conservation is critical to making sure that water supplies remain adequate.

I ♥ SOIL

Janis Boettinger
Soil Scientist
Utah

I was born and raised in New Jersey near New York City. I have always enjoyed the outdoors, especially on family vacations to the West. There I became captivated with the vast, beautiful, yet fragile desert landscapes. I was lucky to have had the opportunity to pursue my graduate education, and ultimately, my career at Utah State University, working in desert landscapes in the West.

Soils are fascinating. Soil is the foundation of all life in terrestrial ecosystems—bridging the biological and physical worlds. To me describing a soil profile is interpreting the record of geological, climatic, ecological, and human events. The integration of biological and physical processes over such vast spans of space and time, especially in deserts, is awe-inspiring.

Temperate forest soils

Some forests have a lot in common with Goldilocks.

If it is not too hot or not too cold—and with just the right amount of moisture—a temperate forest will grow. Their secret is moderation. But not all temperate forests are alike, so neither are their soils.

Under coniferous forests (trees with needles, usually evergreens), the soils are usually sandy and not naturally fertile. Under deciduous forests (where trees shed their leaves in autumn), fallen leaves form a rich upper layer of soil that can be good for farming.

Temperate forests cover about one-third of North America.

TWO COMMON FOREST SOILS

SPODOSOL Greek *spodos* = ashy
ALFISOL refers to the elements *Al* (aluminum) and *Fe* (iron)

Spodosols have a gray or whitish **eluviated** horizon because water leaches minerals and organic materials downward where they form a dark-colored layer. Alfisols are not as leached as Spodosols, but clay from the surface horizons accumulates in the **subsoil**.

ALFISOL

SPODOSOL

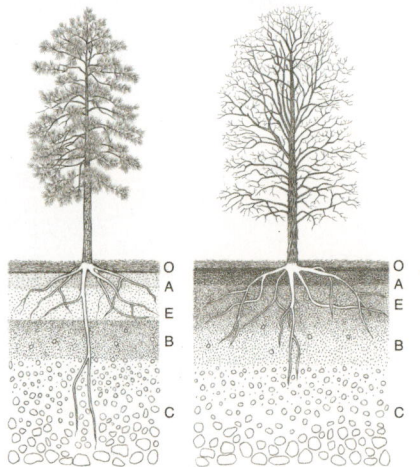

Spodosol forms in a coniferous forest (left), Alfisol in a deciduous forest (right).

FOREST CLORPT

> **CL**imate

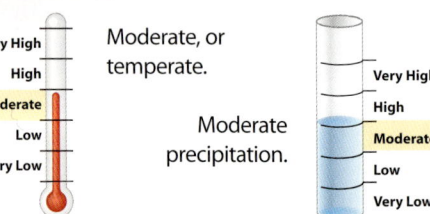

Moderate, or temperate.

Moderate precipitation.

> **O**rganisms

Fungi, bacteria, trees, deer, squirrels, salamanders.

> **R**elief
Varies from flat to steep.

> **P**arent material

Varies.

> **T**ime

Varies. Youngest in areas that had glaciers during the last Ice Age.

Forests Lock in Carbon . . . Keeping It Out of the Air

Temperate forests store a great deal of carbon in their trees and in the soil's organic matter. That keeps carbon from being released into the air and contributing to global warming. But when a forest is cut down to provide lumber or land to build houses and roads, the soils can release their stored carbon. The world's soils store twice as much carbon as the atmosphere.

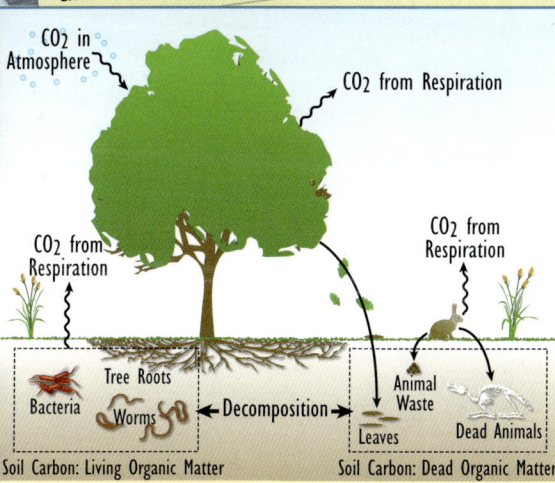

Carbon cycles between the atmosphere, plants (living and dead), and the soil. When plants are removed, the cycle is disrupted.

Clear-cutting (removing all the trees at once) releases carbon to the atmosphere. Forests, however, can be logged for lumber in ways that help to keep carbon in the soil. One practice is to cut down selected trees and leave others. That method also helps to reduce soil erosion by allowing vegetation and organic matter to stay on the surface, which holds soil in place.

When it rains, some water flows downhill into rivers and streams. Some water soaks into the soil, where plants use it. And some water seeps through soils and rocks, where it recharges groundwater. Our drinking water is drawn from streams, rivers, and groundwater.

HEALTHY SOILS, HEALTHY WATER

Do soils in the forest affect the water in a stream? Yes! All soils affect both water quality (how clean the water is) and water quantity (the amount of water) in streams, rivers, and lakes. It even affects groundwater—water that can be hundreds of feet underground.

Every drop of water we drink traveled through soils at one time or another. The soils helped purify it along the way.

Plants keep soils in place so the soil can filter out pathogens (harmful germs), nutrients (good for plants but not for streams or groundwater), and other pollutants.

>>Inside Scoop

The largest forest organism is *not* a tree but a giant fungus that spreads underground in Oregon's Blue Mountains for nearly 6.4 square kilometers (four square miles).

I ❤ SOIL

**Wendy Greenberg
Soil Scientist
Minnesota**

I have been digging in the soil for my whole life, but it was only in college that I discovered that one could actually be a soil scientist. I have been studying and working with soils ever since. I worked on making soil maps in Nepal, as a Peace Corp volunteer, and in different states for the Natural Resources Conservation Service. I studied some very sticky clay soils in Texas and some non-sticky tropical clay soils in Jamaica.

Now I live in northern Minnesota in a small house on loamy sand, which is good for houses but not so good for gardens—it has few nutrients and drains too fast. I teach soil science to college students at Bemidji State University and at the University of Minnesota-Crookston, and sometimes to my two children and their friends at Horace May Elementary School.

The tropics occur in a narrow band 23.5° north and south of the equator (between the Tropic of Cancer and the Tropic of Capricorn). There are tropical regions in Southeast Asia, Central and South America, most of Africa, the Indian subcontinent, parts of Australia, and the Caribbean islands.

Humid tropical soils

A rainforest is lush, *but* its soil is not fertile.

That's because *tropical* soils are highly weathered and low in organic matter. High temperatures year-round and high rainfall, during at least part of the year, break down organic material and minerals very quickly. As a result, little organic matter remains in these soils to make them fertile. Intense rain also leaches nutrients. In a rainforest, trees work in partnership with fungi to quickly absorb nutrients before they are leached to lower soil horizons.

OXISOL French *oxydé* = oxidized
ULTISOL Latin *ultimus* = last

Tropical soils are some of Earth's oldest soils. Being so close to the equator, glaciers did not cover them during the last Ice Age. As a result, they have had a long time to develop and to weather. Iron oxides (rust) give them their reddish brown to yellow color. Ultisols have a layer of clay. Oxisols, however, have no obvious layers.

IT'S NOT ALL JUNGLE

Say tropical, and a rainforest, or jungle, comes to mind. But large parts of the tropics are savannas—rolling grasslands scattered with shrubs and isolated trees. Not enough rain falls there to support forests. The savannas of East and South Africa are probably the most familiar, but there are savannas in South America and Australia as well.

OXISOL

ULTISOL

Can you find where the layer of clay begins?

TROPICAL CLORPT

> CLimate

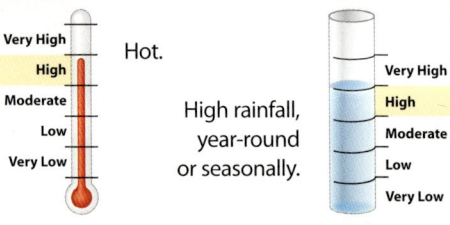

Very High
High — Hot.
Moderate
Low
Very Low

Very High
High — High rainfall, year-round or seasonally.
Moderate
Low
Very Low

> Organisms

Forests: ants, fungi, trees and vines, monkeys, tigers, tree frogs, birds.

Savanna: termites, ants, bacteria, fungi, grasses, acacia trees, elephants, zebras, wildebeests, lions.

> Relief

Varies from flat to rolling to steep mountain slopes.

> Parent material

River sediments or bedrock that has weathered in place.

> Time

Old, highly weathered.

Termites: Tropical Earthmovers

In tropical soils, termites are the chief clean-up crews—feasting on dead wood, leaves, and grasses. Like ants, termites are social insects that live in colonies. Termites build elaborate nests from soil cemented with their saliva. Their mounds can rise up to 9 meters (30 feet). In the process of building, the termites act like a garden tiller, loosening the soil and mixing it with organic matter.

Dig Deeper

Explore how tropical farm land can be managed to keep soil healthy.
www.icajapan.org/virtualtoure/ 01PhilRefE.html

Here a forest was burned to create a pasture. Burning vegetation releases carbon dioxide into the atmosphere, where it can contribute to increased global temperatures.

SOIL CAN TURN TO DIRT

To grow crops, land has to be cleared of native vegetation. But in the tropics, most of the organic matter is tied up in living plants and quickly disappears once the land is cleared. A plot of cleared ground may only be good to farm for a few years before the nutrients run out. The soil becomes poor for crops, native vegetation, and people: The soil turns to "dirt." As a result, farmers must clear more and more land to raise their crops.

There are no easy answers to this dilemma. But proper farming methods and soil management can make it possible to restore the soils and reduce the need to clear as much land.

Without trees, tropical soils quickly dry out and become desert-like. They also easily erode without vegetation and organic matter to hold the soil in place.

I ❤ SOIL

**Monday Mbila
Environmental Soil Scientist
Alabama**

My earliest memory of soils is related to the ease or difficulty with which we prepared the soils for planting in our family farms. But, my interest in soil science as a scientific pursuit came later as a college student in agricultural science. That interest had something to do with the fact that my favorite professors were all soil scientists. Since then I have worked with tropical and subtropical soils from different parts of the world to better understand the impacts of people on soils.

Soil science is very interesting because soil interfaces with all aspects of the environment—geology, biology, hydrology, and the atmosphere. Soil science holds the key to solving many of today's problems such as managing greenhouse gases and climate change, water resources, and providing food for the world's increasing population. I currently teach soil science to undergraduate and graduate students at Alabama Agricultural and Mechanical University.

Organic soils form in low, wet areas where decayed plants accumulate. Because wet soil has little or no oxygen, dead plants decompose very slowly. Over time, a thick layer of organic material builds up to form organic soils.

wetland soils

In its simplest definition, a wetland is *wet* land.

Wetlands are found everywhere, from the tropics to the tundra, in grasslands or at the seashore. They even occur in deserts. Wetland soils often form in flat, low-lying areas or in depressions where water from rain or snow collects. They may also develop near springs where water is near, or above, the surface. But wherever a wetland forms, the soil stays wet (saturated) because it does not drain well.

Wetlands are important habitats for wildlife from fish to frogs to flamingos. They protect against floods and help clean out pollutants in **estuaries**, rivers, and lakes.

HYDRIC SOIL Latin *hydros* = water

All wetland soils share common colors and color patterns. The surface layer is often black because organic material accumulates there. The subsoil is gray with mottles of bright oranges and reds where iron has oxidized, or rusted. Some very wet soils may even be blue, green, or purple.

HYDRIC SOIL

HISTOSOL

Histos is Greek for organic. These wetland soils have a very thick layer of black organic matter from decomposed plants.

NATURE'S NURSERIES

More than 200 species of birds nest and raise their young in or near wetlands. For this mother duck, the maze of wetland plants makes a good hiding place for her young, as well as a source of food and nesting materials. Wetlands are important nurseries for many other animals, ranging from insects to fish to alligators.

When drained, organic soils can make good farmland if they are properly managed.

> **CL**imate

Any.

Very little to a lot of precipitation, as long as the soil remains wet.

Very High / High / Moderate / Low / Very Low

> **O**rganisms

Anaerobic bacteria, insects, wetland plants (rushes, reeds), water loving trees (Cypress, swamp maple, mangrove), beaver, waterfowl.

> **R**elief

Depressions, flat land. Poorly drained.

> **P**arent material

Varies.

> **T**ime

From very young to very old.

Preserved in Peat . . . Bog People

In 1950, two Danish men were cutting **peat** for fuel when they saw a human face poking through the soil. They thought they had found a recent murder victim, as the body's skin and flesh were intact. But a local **archaeologist** determined that the body was about 2000 years old. But how was that possible? In most soils, skin and flesh decompose quickly. Peat, however, has special properties that slow the bacteria that decompose dead organisms.

Hundreds of well-preserved bodies have been found in peat bogs, providing clues about life in the ancient past.

Organic soils known as peat are dug up and dried for fuel. Peat burns because it has a very high carbon content like charcoal or coal.

This is a drained wetland in Florida. The post marks the height of the soil about 80 years ago. The soil has sunk about two centimeters per year (one inch), which is faster than the organic soil can form.

The loss of a wetland also means a loss of habitat for wildlife.

DRAINED WETLAND WOES

Wetlands can help protect people and property from floods. That's because they soak up and hold water like a sponge. During heavy rains, the water spreads out and slowly soaks into wetland soils instead of rapidly running into rivers and streams, which may cause them to flood. But when wetlands are drained and replaced by farms and housing, or drowned by rising sea level, the risks of flooding increase.

Draining organic wetland soils adds another problem. As these soils dry, they shrink. Think of a wet sponge versus a dry sponge. As a result, they do not make very stable platforms to build on.

The loss of wetlands may have contributed to the flooding of New Orleans during Hurricane Katrina in 2005.

I ♥ SOIL

**David Lindbo
Environmental Pedologist
North Carolina**

Although I grew up in the suburbs of Boston, Massachusetts, my parents and grandparents had a strong connection to farming and the land. My first experience with soils was in elementary school working with my grandfather in his vegetable garden. He wanted his grandchildren to understand where food came from and how to grow it without hurting the environment. I only recall using pesticides a few times. And we used manure and compost instead of commercial fertilizer whenever possible.

In college, one of my professors rekindled my interest in soils. He showed how life (as we know it) couldn't exist without soils. When I moved on to graduate school, I knew soils were my field. Today as a soil scientist, I focus on how to use soils to treat **sewage**—*a method used by over 25% of U.S. homes. As this percentage increases in the future, soil will be doing even more for us.*

the big

Soil connects it all.

picture

clean because of soil

air

grown in soil

food

lives in soil

insects

nourished by soil

people

water

filtered by soil

lives in trees growing in soil

animals

rock

shelter

built on soil from trees grown in soil

proud parent of soil